FARMALL CUB

PHOTO ARCHIVE

IN MEMORIAM

PETER LETOURNEAU
1950–1997

We dedicate this book as a memorial to our friend, partner, and fellow worker Peter Letourneau, who died driving a 1926 3 litre Bentley, a passion close to his heart. We will miss him and wish him God Speed.

Shawn Glidden, Theresa Piemonte, Rick and Debbie Seymour, Tom Warth

FARMALL CUB
PHOTO ARCHIVE

Photographs from the
McCormick-International Harvester Company Collection

Edited with introduction by
P.A. Letourneau

Iconografix
Photo Archive Series

Iconografix
PO Box 446
Hudson, Wisconsin 54016 USA

Books in the Iconografix *Photo Archive Series* are offered at a discount when sold in quantity for promotional use. Businesses or organizations seeking details should write to the Marketing Department, Iconografix, at the above address.

Library of Congress Card Number 97-70639

ISBN 1-882256-71-9

97 98 99 00 01 02 03 5 4 3 2 1

Printed in the United States of America

PREFACE

The histories of machines and mechanical gadgets are contained in the books, journals, correspondence, and personal papers stored in libraries and archives throughout the world. Written in tens of languages, covering thousands of subjects, the stories are recorded in millions of words.

Words are powerful. Yet, the impact of a single image, a photograph or an illustration, often relates more than dozens of pages of text. Fortunately, many of the libraries and archives that house the words also preserve the images.

In the *Photo Archive Series*, Iconografix reproduces photographs and illustrations selected from public and private collections. The images are chosen to tell a story—to capture the character of their subject. Reproduced as found, they are accompanied by the captions made available by the archive.

The Iconografix *Photo Archive Series* is dedicated to young and old alike, the enthusiast, the collector and anyone who, like us, is fascinated by "things" mechanical.

Iconografix continuously seeks collections of archival photographs for reproduction in future books. We require a minimum of 120 photographs per subject. We prefer subjects narrow in focus, i.e., a specific model, railroad, racing venue, etc. Photographs must be of high-quality, suited to reproduction in an 8x10-inch format. We willingly pay for the use of photographs.

If you own or know of such a collection, please contact: The Publisher, Iconografix, PO Box 446, Hudson, Wisconsin 54016.

The photographs and illustrations that appear in this book were made available by the State Historical Society of Wisconsin. The Society is the official repository for records of the International Harvester Company and its nineteenth-century predecessor, the McCormick Harvesting Machine Company.

The McCormick-International Harvester Company Collection contains nearly 4,000 cubic feet of family papers and business records, including technical publications, advertising literature, engineering and promotional photographs, posters, and films. Its cataloguing, preservation, and administration are funded through an endowment established in 1991 by Brooks McCormick.

Use of the collection is not generally restricted. However, due to its size and complexity, interested persons are encouraged to contact the Society in advance, at 816 State Street, Madison, Wisconsin 53706.

International Harvester's Farmall Cub, introduced in 1947 at $545, "designed and built to do easier, faster, and better, every job ordinarily handled by two or three horses or mules..."

INTRODUCTION

The original Farmall began life as a motorized cultivator, and, following a decade of engineering, evolved into what became the first all-purpose tractor. With its adjustable wheel tread, ample crop clearance, and optional power take-off, the Farmall Regular of 1924 could be used to not only plant but also cultivate and harvest row crops—jobs that standard tractors simply could not do. Within five years of the Farmall's introduction, almost every major tractor manufacturer offered a comparable tractor, and by 1940 general purpose tractors comprised 90 percent of the tractors built in the United States.

In the 25 years that followed its arrival, IH introduced larger and smaller versions of the Farmall, such that by 1940 there was a Farmall suited to almost any size farm. In a sense, the Cub was the "final" Farmall. Introduced in 1947, the Cub was designed to meet the needs of the few remaining farmers for whom the price of a new tractor was still not a practical expenditure—truck gardeners, nursery men, landscapers, and any remaining farmers who still relied on draft animals. In addition, IH promoted the tractor for auxiliary use on larger farms. A Cub offered all the benefits of larger Farmalls. With its controls and seat offset to the right side of the tractor, the Cub offered a nearly unobstructed forward view. According to IH, "This clear vision feature, which we call 'culti-vision,' permits you to see all front-mounted implements at work as well as the row ahead. You can travel swiftly down the row... even a row of tiny seedlings... and do a close-up job."

The Cub changed little during the course of its production. It utilized an IH, 4-cylinder, 59.5-cubic inch, L-head engine that produced 9 ¼ belt horsepower, in the early Cub, and 10 ¾ belt horsepower in later versions. Throughout production, Cub was equipped with a 3-speed transmission. Early tractors offered magneto ignition as standard equipment. Electric starting was optional but became a standard feature in 1958. "Touch-Control" hydraulics, originally an option, became standard equipment by 1955. "Fast-Hitch", IH's response to the Ferguson three-point hitch, was also a standard feature of later machines. Power take-off, belt pulley, swinging drawbar, adjustable front axle, lights, and wheel weights were among the few remaining items offered as options.

Over 215,000 Farmall Cub tractors were built between 1947 and 1964, with better than 80 percent of the total production built by the end of 1953. The Farmall Cub was successful as much because of the variety of Cub-sized implements that IH offered as because of the tractor itself. Plows, planters, cultivators, mowers, and more were specially designed to match the size and work output of the Cub. There was little that the tractor could not do. The photographs in this volume are testimony to the words of International Harvester, which described this smallest of Farmalls as "a Cub in size but a bear for work."

Farmall Cub Photo Archive presents photographs selected from the McCormick-International Harvester Company Collection. The photographs, for the most part, were taken between 1945 and 1951, the Cub's early years. Nonetheless, as the Cub underwent few significant changes through the course of its production, the book successfully captures the very spirit of the Farmall Cub, and ably conveys the variety of tasks that this rugged and still popular tractor was called upon to handle.

The first Cub tractor, produced at the Louisville Works, Louisville, Kentucky.

From: INTERNATIONAL HARVESTER COMPANY
CHICAGO, ILLINOIS

FOR RELEASE OCTOBER 15, 1945

Photograph I-3158-II

Now Little Farmall Tractor for Small Acreages - One of the most important additions to International Harvester Company's line of farm machines for postwar is the new Farmall Cub, designed to meet the power needs of farmers with small acreages. This tractor, now in the preproduction stage, is intended to offer the small-acreage farmer the many advantages of low-cost mechanized farming enjoyed by those operating larger farms. The first public showing of the Farmall Cub and equipment was made by Harvester recently as the highlight of its large field demonstration of farm machines for postwar at Hinsdale, Illinois. The Company announces that a complete line of attachable machines will be made available for the Farmall Cub when it goes on the market. Authorities on farm machines consider the introduction of this tractor one of the most outstanding developments in agricultural equipment in many years. The new little tractor is shown here operating a one-furrow quick-attachable moldboard plow.

Promotion of the Cub began far in advance of actual production. This photograph, offered to the farm press in October 1945, was accompanied by the press release appearing on the opposite page.

Cub with optional electric starter and lighting device and optional Touch-Control hydraulics.

Cub with magneto ignition and optional Touch-Control hydraulics. The offset design of the Cub permitted outstanding forward vision. IH called this clear-vision feature "culti-vision".

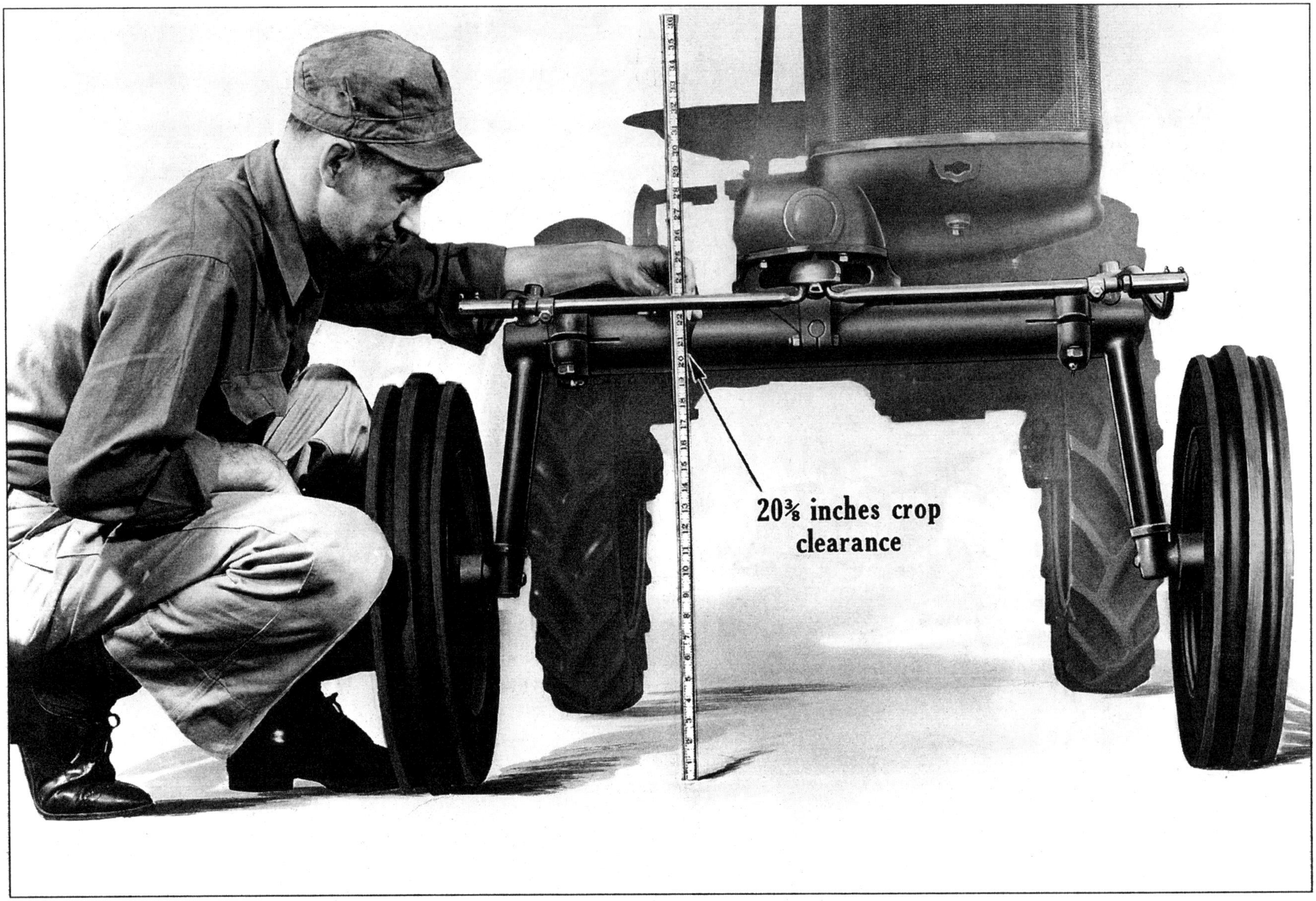

Cub offered crop clearance of 20⅜ inches for the full width of the front axle. With "regular" front axle, front wheel tread spacing could be set at 40⅝ inches or 46⅜ inches by simply reversing the wheels on the hubs.

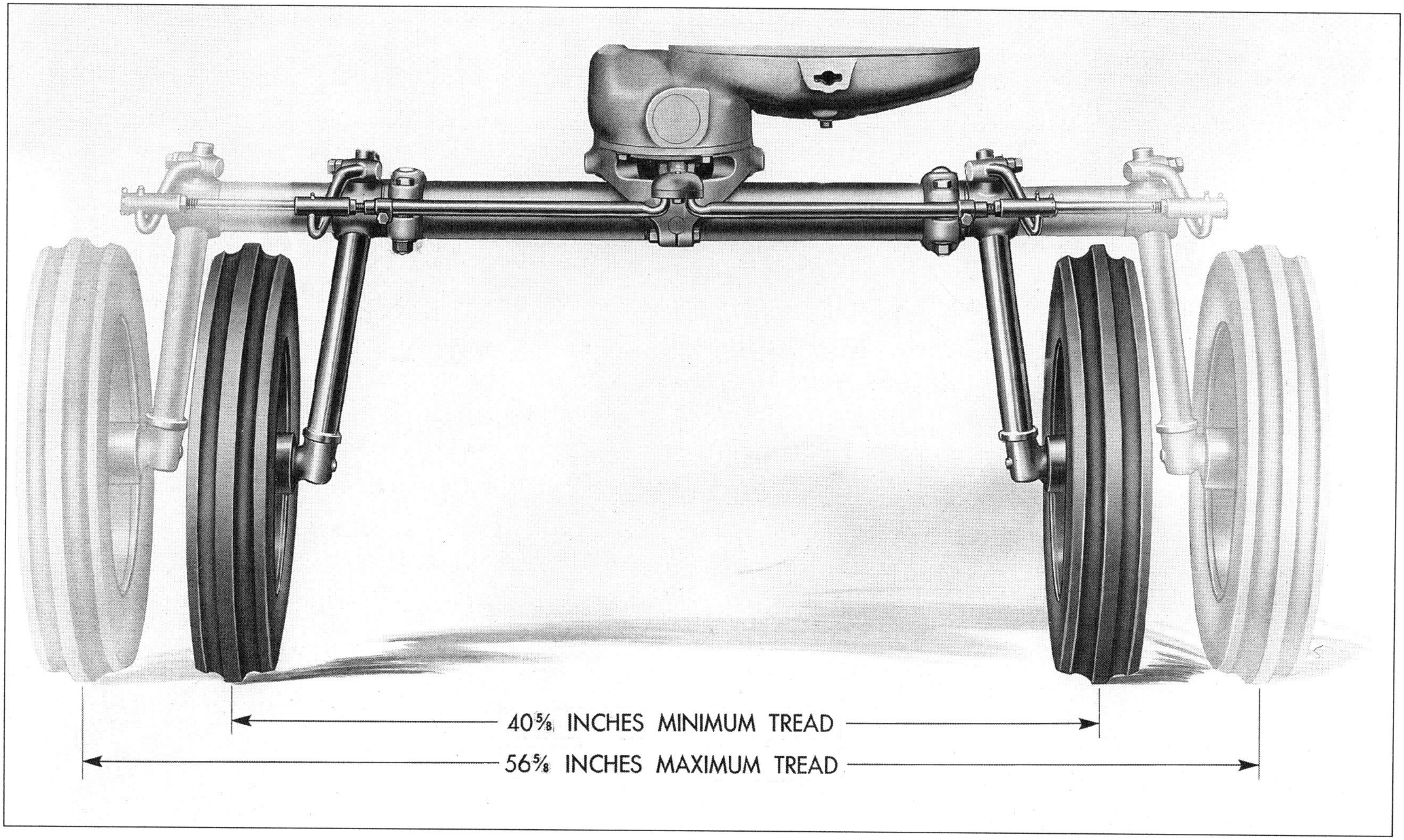

With optional adjustable-tread front axle, the Cub's front wheel tread could be matched to that of the rear, and was variable from 40⅝ inches to 56⅝ inches in four-inch increments.

The Cub's rear wheel tread was adjustable from 40 to 56 inches in four-inch increments.

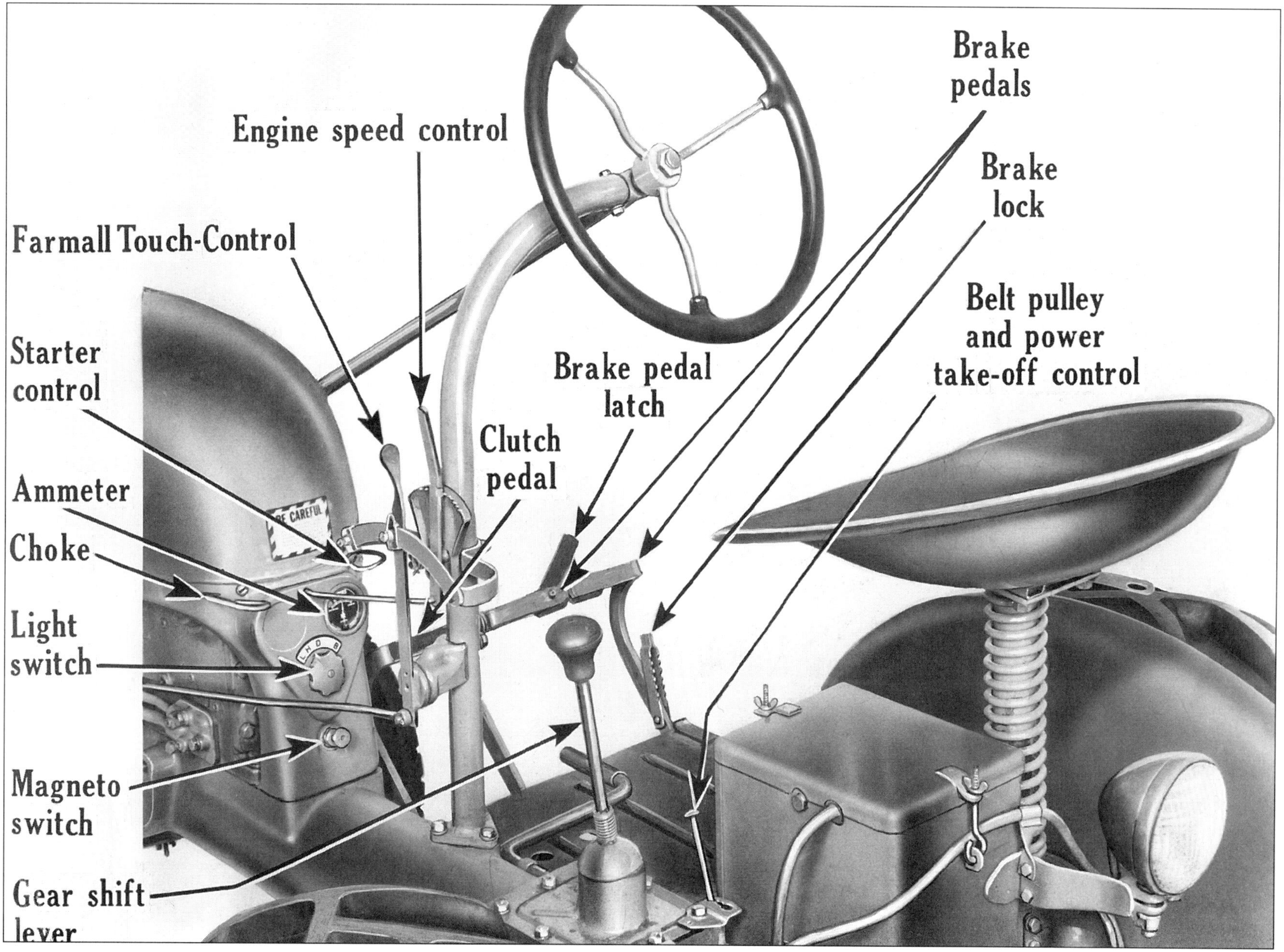

A retouched photograph of the Cub's operator platform and controls, used in brochures to illustrate the Cub's level of operator comfort and ease of finger tip control.

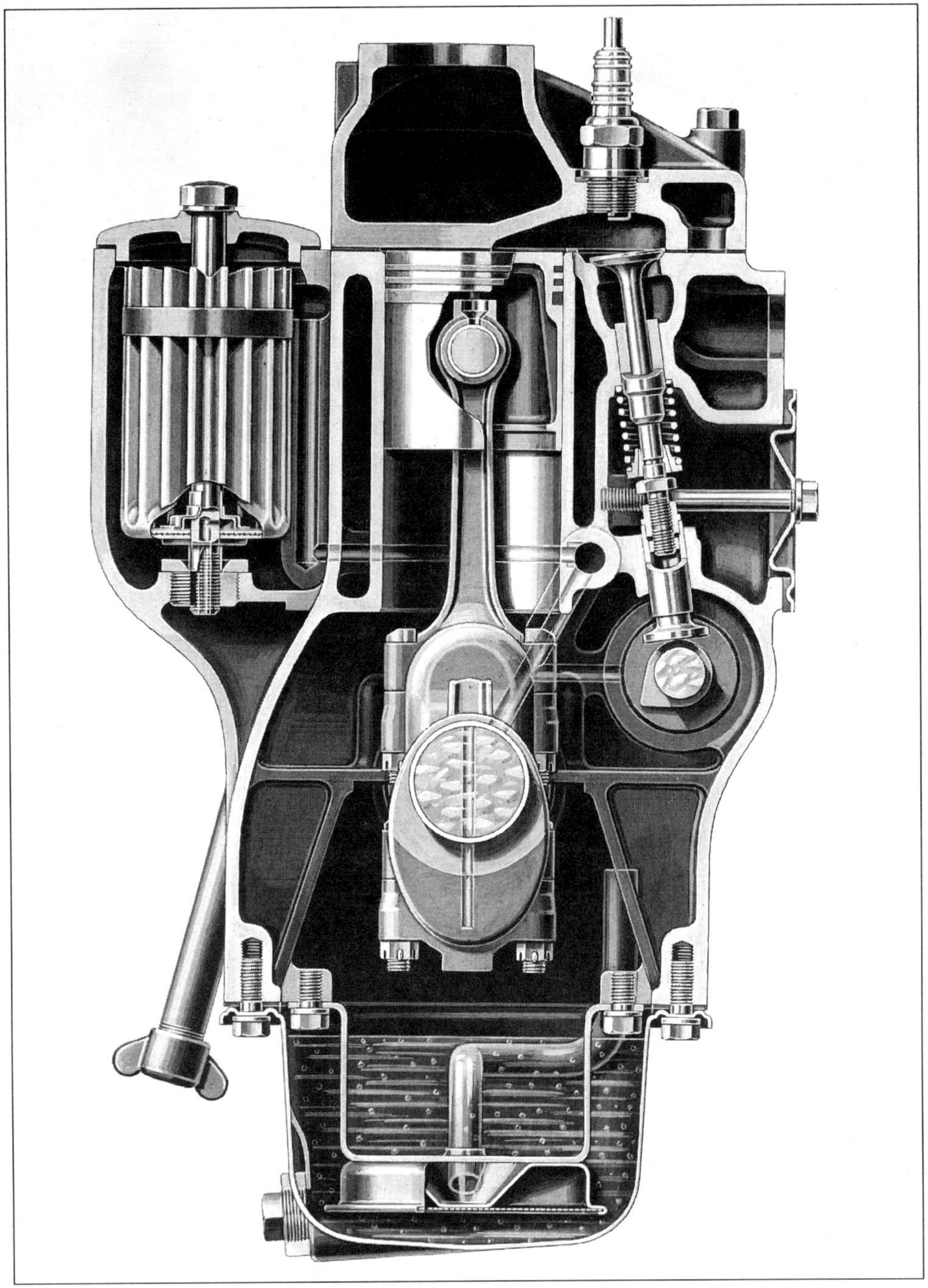

Cutaway illustration of the Cub four-cylinder, 59.5-cubic inch gasoline engine.

The Cub's replaceable paper element oil filter removed particles as small as a micron. International recommended oil changes every 120 hours.

Throughout much of production, the Cub was offered with an IH Model J-4 magneto ignition as basic equipment.

The early Cub's optional electric starter and lighting device.

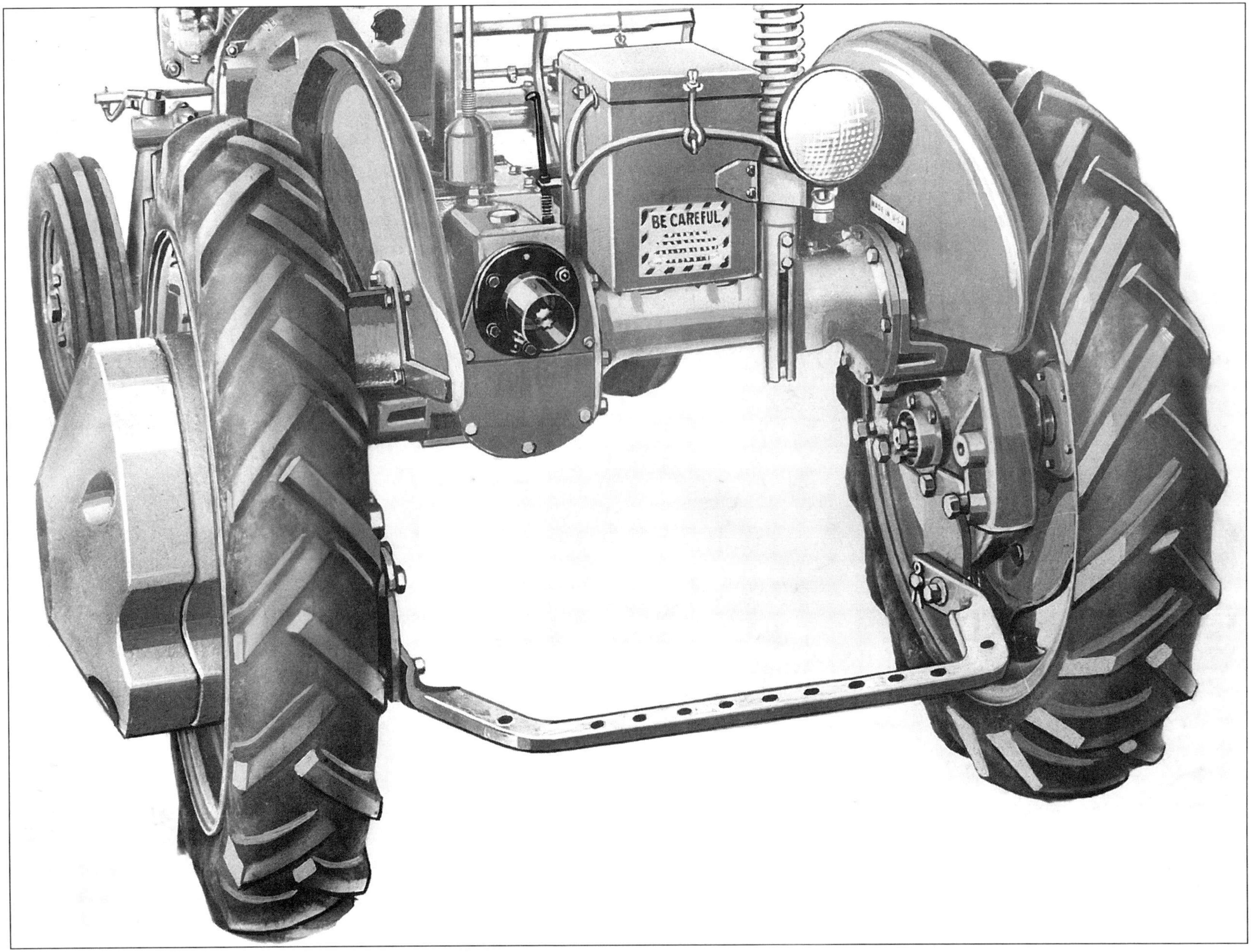

Rear views showing the optional 1600-rpm power take-off (above) and PTO-driven belt pulley attachment (opposite page).

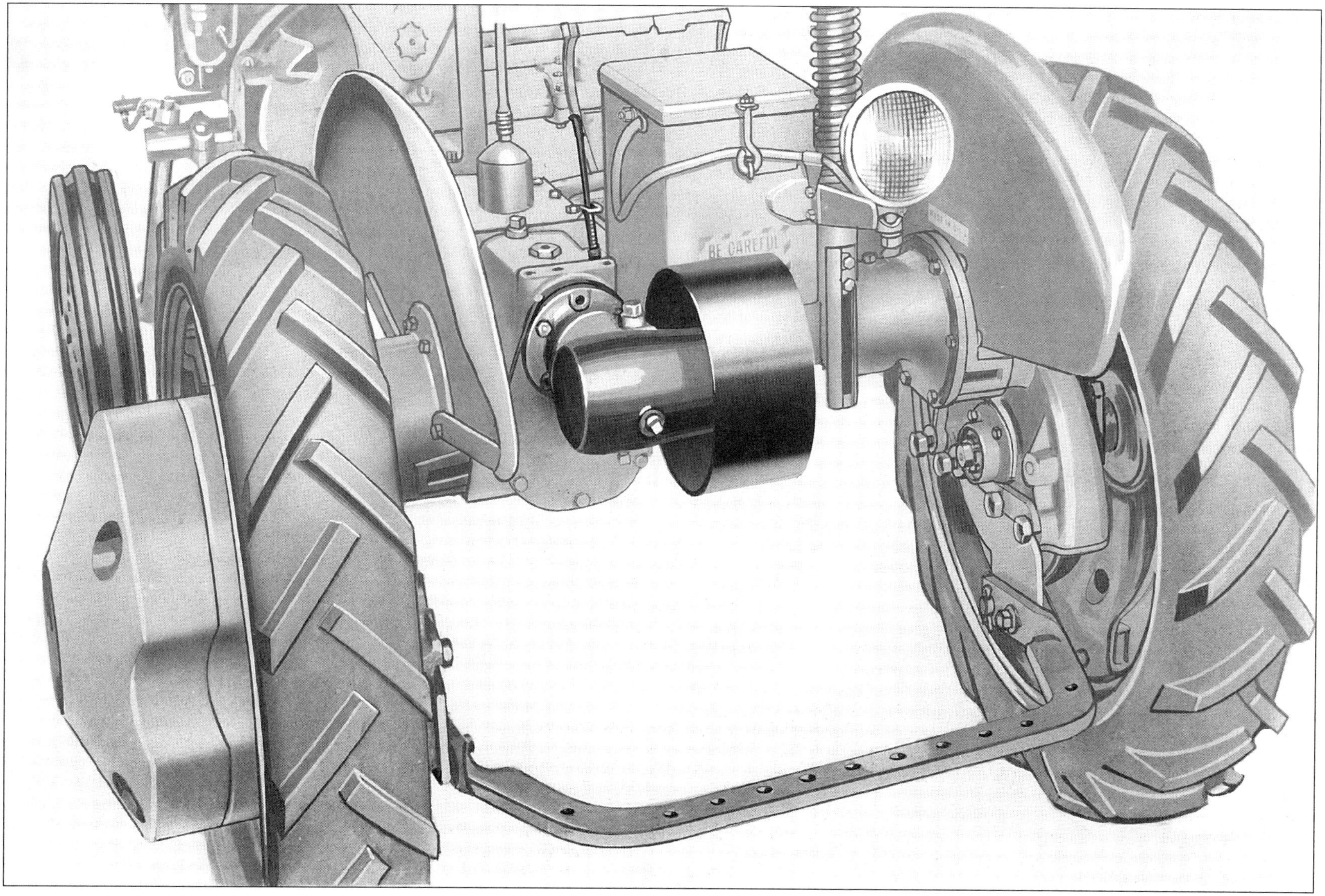
BE CAREFUL

The Cub's optional hydraulic Touch-Control system became basic equipment by the mid-1950's.

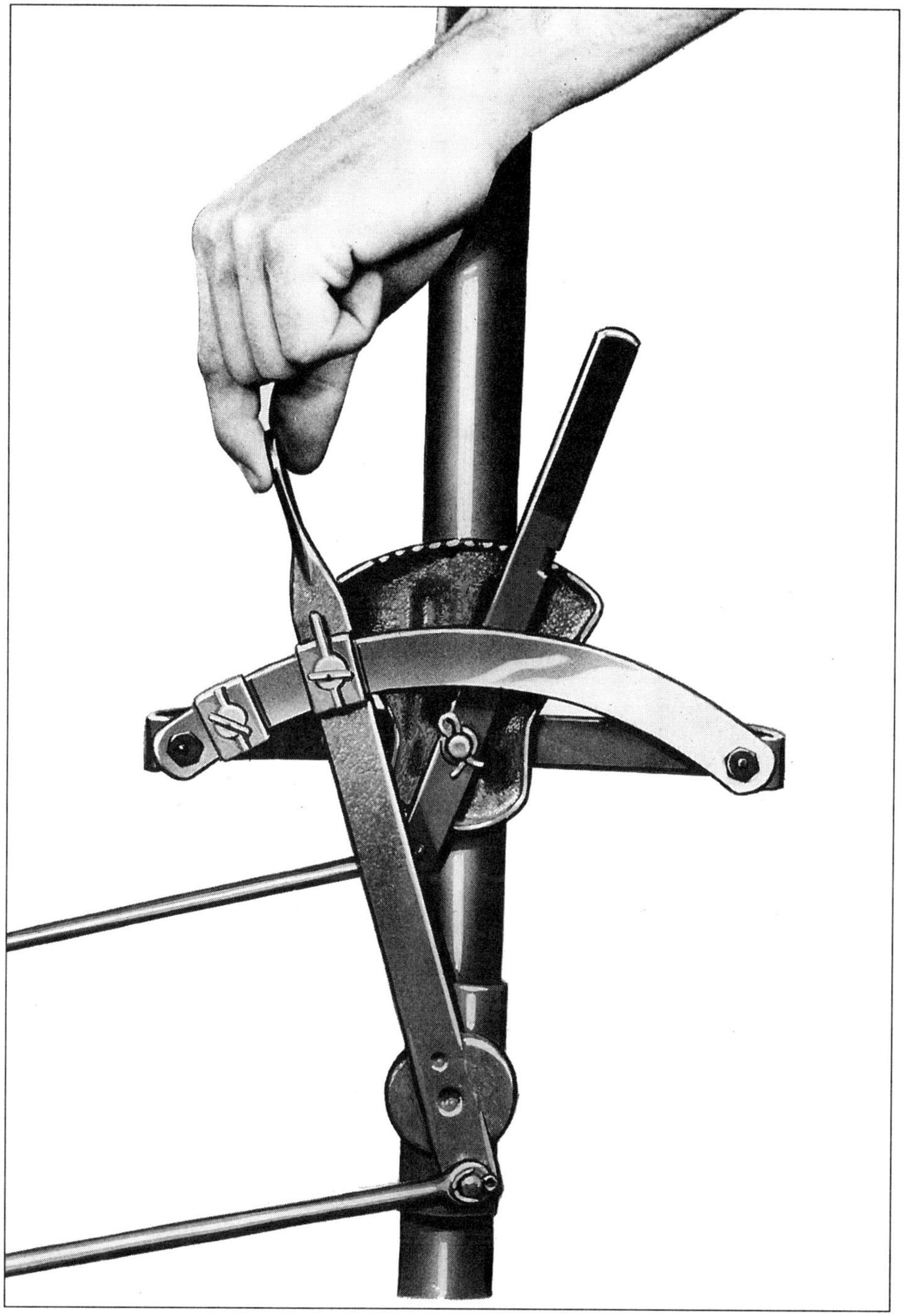

Control mechanism for Touch-Control hydraulics. (Smaller lever was the hand throttle.)

Optional 30-lb. front wheel weight.

Optional 145-lb. rear wheel weight.

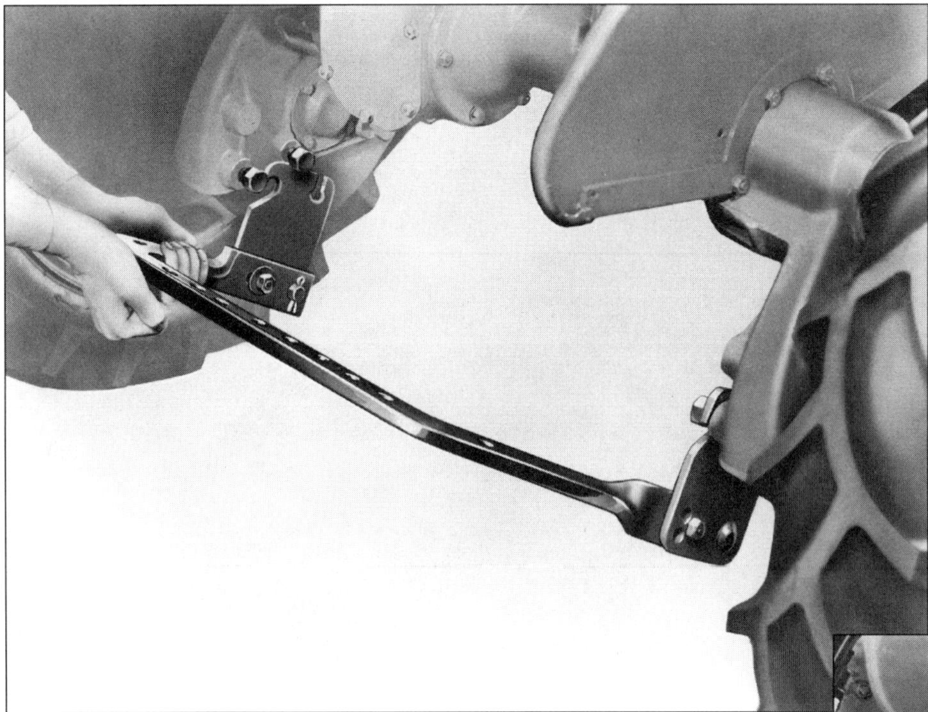

Removing drawbar, preparatory to attaching rear-mounted implement.

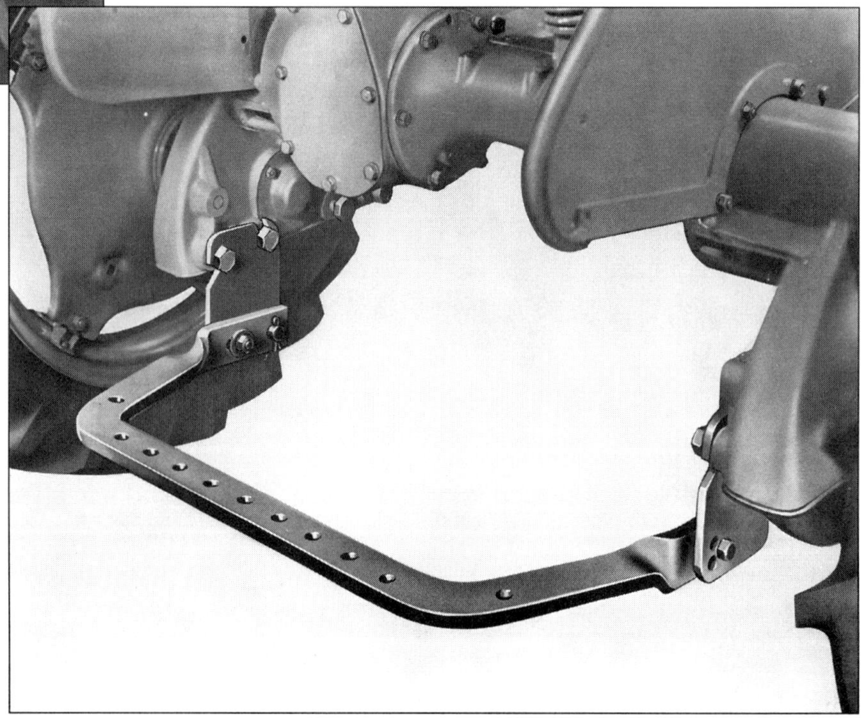

Reversible floating drawbar (basic equipment) in center position, set for hitching trail-behind implements.

Mounting an implement on the rear of the Cub was an easy matter. Shown is Mr. Farmer slipping the rear section of the Cub-144 cultivator onto the rear Tractor Mounting Pads. Two bolts held each rear section in place.

Rear view of a Cub illustrating the optional swinging drawbar.

Cutaway illustration of the Farmall Cub.

The Cub was offered with a variety of "scaled-to-size" implements. A pre-production tractor with direct-connected (mounted) one-row cultivator, and selected pre-production implements posed for the camera. Implements included: four-foot tandem disk harrow, mower, Southeast-type planter with fertilizer attachment, vegetable planter, vegetable cultivator, two-way plow, Southwest-type cotton and corn planter, and one-furrow disk.

The Cub was promoted as "the ideal, low-cost, all-purpose tractor for the small farm... truck garden... or for auxiliary use on large farms." Here a Cub was featured in a window display of a Buffalo, New York men's store, in an effort to reach the suburban gardener or small estate holder.

Preparing a rail shipment of new Cub tractors, 1947.

A rail shipment of Cubs from the new Louisville Works, destined for shipment to the Southwest District branch.

A lineup and parade of new Farmall Cub tractors were common promotional events. These photos were taken at a public demonstration held at the Dominion Government Central Experimental Farm, Ottawa, Canada.

Cub FARMALL
WITH

County and state fairs were occasions for marketing the Cub and its various "scaled-to-size" implements and attachments. Here is a Cub with the 4½-foot, PTO-driven Cub-22 mower at the Illinois State Fair.

Cub with Cub-151 one-furrow disk plow at a field introduction. Promotional photos often featured women, children, and non-farming "city" folk, to illustrate the ease with which the Cub was operated.

Dealer-sponsored farm shows were another means of promoting the new Cub. Here two Cubs (and Farmall M to the right) were featured at such a show in Sweetwater, Texas.

A Cub matches size and power with a Frisco Lines diesel locomotive, in this publicity photograph. Springfield, Missouri, 1947.

Cub with Cub-193 one-way plow.

Cub with Cub-193 plow. The Cub-193 plow was a one-furrow moldboard plow with a 12-inch bottom and a right-hand beam. A 16-inch plain rolling colter was standard.

Pre-production Cub, with pre-production (Cub-151) disk plow, plowing on the contour.

Pre-production Cub with pre-production (Cub-151) single-furrow disk plow.

Cub channel terracing with a lister plow on the Avery Community Farm, Canton, Georgia, 1949.

Cub with Touch-Control hydraulics on a terracing job.

Two views of a pre-production Cub with pre-production (Cub-23A) tandem disk harrow. November 1945.

FARMALL

Cub with two-section peg-tooth harrow.

Cub with 10-foot, no. 6 weeder mulcher attachment for Cub-144 cultivator.

Cub with 54-inch, Cub-3 spring-tooth field cultivator on grassy land that was to be planted in spinach.

A farmer gathers rocks from his field before drilling wheat. The Cub (shown with McCormick trailer) was used for multiple tasks on this 700-acre farm.

A Santa Monica, California part-time farmer and his son on their 10-acre ranch. The pair fed 100 to 125 hogs a year and kept one cow. In addition to the tractor, they bought a Cub plow, cultivator, disk harrow (shown), spring-tooth cultivator, mower, and leveling blade (shown).

Cub with grading and leveling blade filling in roadway to the farmhouse. Lafayette, Indiana, April 1947.

Cub with grading and leveling blade cleaning off concrete hog feeding area. Lafayette, Indiana, April 1947.

Cub with grading and leveling blade maintaining roads on International Harvester's Hinsdale Farm.

FARMALL C

Cub with specially made, hydraulically operated blade at work on Wilson Park Housing Project, Milwaukee, Wisconsin, 1948.

A Cub used to plow snow to permit coal deliveries in Manistique, Michigan.

Cemetery caretaker digging for a monument base. He used the Cub to haul away dirt. Valpariso, Illinois, April 1947.

Bringing up sand for fill on road shoulders using Cub with grading and leveling blade. This non-farmer owner lived on a rural home site near Chesterton, Iowa, 1950.

Cub with pre-production vegetable planter.

Cub with Cub-474 vegetable planter with marker.

Two views of a Cub with Cub-474 four-row vegetable planter.

FARMA

Cub with grain drill sowing and fertilizing wheat.

Cub with Cub-172 one-row corn and cotton planter.

Cub with a Cub-172A one-row runner planter with fertilizer attachment.

Cub with Cub-172 corn and cotton planter planting baby lima beans.

Illustration of the hopper drive chain connection to drive sprocket on rear axle, used with Cub-170 to 172 planters.

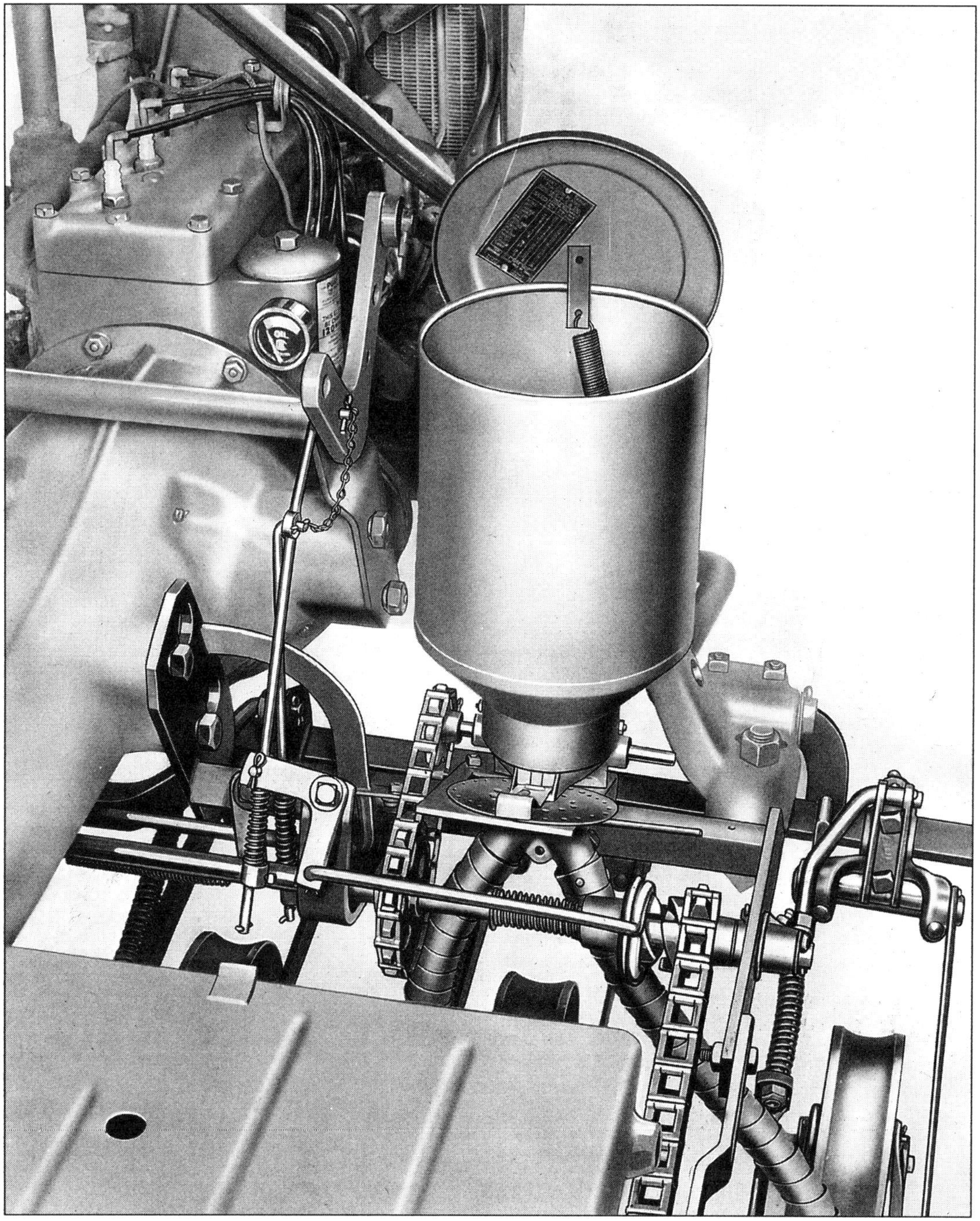

Clutch detail on the Cub-474 planter.

A Cub used to tow airplanes at the Elmhurst, Illinois airport. May 1949.

Special tow bar and bracket fitted to the Cub handling airplanes on the opposite page.

A Cub with a model 200 McCormick spreader passes a Farmall MD with disc harrow.

Farmall Cub with McCormick wagon.

Cub snaking timber out of a wood lot. Sparta, Wisconsin, December 1948.

Cub hauling home logs from a wood lot. Sparta, Wisconsin, December 1948.

Cub operating saw for cutting firewood. Sparta, Wisconsin, December 1948.

A Humboldt, Iowa farmer dragging the lane with his Cub.

Pre-production Cub with pre-production direct-connected mower, October 1, 1945.

Cub with Cub-22 side-mounted mower. The farmer is filling the Cub's 7 ½ gallon fuel tank.

Cub mowing alfalfa hay with a Cub-22 side-mounted mower.

Cub with Cub side-delivery rake in second cutting of alfalfa. This experimental Cub-sized rake was never produced, most likely because the Cub could handle a standard-sized side-delivery rake.

Cub with a standard-sized side-delivery rake in an Illinois hay field.

Pre-production Cub with hay loader. Swanton, Ohio, 1946.

Two views of a Cub with McCormick 100 spreader.

A Cub used to transport seed and supplies on a Texas cotton farm.

Pre-production Cub with wagon, October 1945.

Part-time farmer and full-time section foreman for the North Western Railroad plants corn with his Cub. During summers his hard-working Cub was used to maintain the railroad right-of-way. His son, on the opposite page, pulls a peg-tooth harrow with the Cub. Westside, Iowa, May 1949.

McCORMICK
FARMALL CUB

Cub pulls the hay rope.

Cub moving a brooder house.

Pre-production Cub with single-row cultivator, October 15, 1945.

Pre-production Cub with pre-production direct-connected one-row cultivator in raised position. November 1945.

Two views of a Cub with Cub-447 four-row vegetable cultivator in three rows of beets.

Cub with independent unit cultivator in 14-inch rows. June 1946.

The Cub's offset construction permitted an unobstructed view of the cultivator at work.

Two views of a pre-production Cub with Touch-Control and cultivator with fertilizer attachment in corn. Chula Vista, California.

FARMALL

Two views of a Cub with Cub-144 cultivator in tomatoes. Arlington Heights, Illinois.

Cub cultivating corn.

A pre-production Cub with Cub-144 one-row cultivator and no. 53 ground tool combination in eggplant.

Cub with Cub-144 cultivator in bean field on 3-acre truck garden, Solana Beach, California.

Cub cultivating on a 900-acre Fresno, California farm that grew tomatoes, watermelon, cantaloupe, eggplant, potatoes, lettuce, and carrots.

Cub cultivating corn.

Cub with Cub-144 cultivator in a 100-acre cabbage field near Seneca Castle, New York. This crop was grown for Campbell Soup Company.

A motorized corn detasseling outfit moves through a cornfield...

...and at the end of the rows the detasseling outfit is revealed as...

...a much-modified Farmall Cub tractor. This custom unit was built in Manning, Iowa, August 1947.

Cub operating belt-driven corn sheller.

Cub with an all-steel box hauling a load of corn from the corn shredder.

A farmer grinds oats with a Cub-powered McCormick no. 4E hammer mill.

Cub and a no. 52R combine harvesting wheat near Bozeman, Montana.

An FFA "Star" farmer from the Ozarks used his Cub to power a grinder.

Cub with flame cultivator.

Cub with ammonia fertilizer applicators, 1949.

A Cub with experimental electric Touch-Control lift, 1946.

A wide-tread Cub with six-row planter, 1950.

Late-style Farmall Cub offered standard Touch-Control hydraulics, starter and lights. Featured here with optional deluxe cushion seat, 1957. The Farmall Cub went out of production in 1964.

The Iconografix Photo Archive Series includes:

AMERICAN CULTURE

AMERICAN SERVICE STATIONS 1935-1943	ISBN 1-882256-27-1
COCA-COLA: A HISTORY IN PHOTOGRAPHS 1930-1969	ISBN 1-882256-46-8
COCA-COLA: ITS VEHICLES IN PHOTOGRAPHS 1930-1969	ISBN 1-882256-47-6
PHILLIPS 66 1945-1954	ISBN 1-882256-42-5

AUTOMOTIVE

FERRARI PININFARINA 1952-1996	ISBN 1-882256-65-4
GT40	ISBN 1-882256-64-6
IMPERIAL 1955-1963	ISBN 1-882256-22-0
IMPERIAL 1964-1968	ISBN 1-882256-23-9
LE MANS 1950: THE BRIGGS CUNNINGHAM CAMPAIGN	ISBN 1-882256-21-2
LINCOLN MOTOR CARS 1920-1942	ISBN 1-882256-57-3
LINCOLN MOTOR CARS 1946-1960	ISBN 1-882256-58-1
MG 1945-1964	ISBN 1-882256-52-2
MG 1965-1980	ISBN 1-882256-53-0
PACKARD MOTOR CARS 1935-1942	ISBN 1-882256-44-1
PACKARD MOTOR CARS 1946-1958	ISBN 1-882256-45-X
SEBRING 12-HOUR RACE 1970	ISBN 1-882256-20-4
STUDEBAKER 1933-1942	ISBN 1-882256-24-7
STUDEBAKER 1946-1958	ISBN 1-882256-25-5
VANDERBILT CUP RACE 1936 & 1937	ISBN 1-882256-66-2

TRACTORS AND CONSTRUCTION EQUIPMENT

CASE TRACTORS 1912-1959	ISBN 1-882256-32-8
CATERPILLAR MILITARY TRACTORS VOLUME 1	ISBN 1-882256-16-6
CATERPILLAR MILITARY TRACTORS VOLUME 2	ISBN 1-882256-17-4
CATERPILLAR SIXTY	ISBN 1-882256-05-0
CLETRAC AND OLIVER CRAWLERS	ISBN 1-882256-43-3
ERIE SHOVEL	ISBN 1-882256-69-7
FARMALL CUB	ISBN 1-882256-71-9
FARMALL F–SERIES	ISBN 1-882256-02-6
FARMALL MODEL H	ISBN 1-882256-03-4
FARMALL MODEL M	ISBN 1-882256-15-8
FARMALL REGULAR	ISBN 1-882256-14-X
FARMALL SUPER SERIES	ISBN 1-882256-49-2
FORDSON 1917-1928	ISBN 1-882256-33-6
HART-PARR	ISBN 1-882256-08-5
HOLT TRACTORS	ISBN 1-882256-10-7
INTERNATIONAL TRACTRACTOR	ISBN 1-882256-48-4
INTERNATIONAL TD CRAWLERS 1933-1962	ISBN 1-882256-72-7
JOHN DEERE MODEL A	ISBN 1-882256-12-3
JOHN DEERE MODEL B	ISBN 1-882256-01-8
JOHN DEERE MODEL D	ISBN 1-882256-00-X
JOHN DEERE 30 SERIES	ISBN 1-882256-13-1
MINNEAPOLIS-MOLINE U-SERIES	ISBN 1-882256-07-7
OLIVER TRACTORS	ISBN 1-882256-09-3
RUSSELL GRADERS	ISBN 1-882256-11-5
TWIN CITY TRACTOR	ISBN 1-882256-06-9

RAILWAYS

CHICAGO, ST. PAUL, MINNEAPOLIS & OMAHA RAILWAY 1880-1940	ISBN 1-882256-67-0
CHICAGO&NORTH WESTERN RAILWAY 1975-1995	ISBN 1-882256-76-X
GREAT NORTHERN RAILWAY 1945-1970	ISBN 1-882256-56-5
MILWAUKEE ROAD 1850-1960	ISBN 1-882256-61-1
SOO LINE 1975-1992	ISBN 1-882256-68-9
WISCONSIN CENTRAL LIMITED 1987-1996	ISBN 1-882256-75-1

TRUCKS

BEVERAGE TRUCKS 1910-1975	ISBN 1-882256-60-3
*BROCKWAY TRUCKS 1948-1961**	ISBN 1-882256-55-7
DODGE TRUCKS 1929-1947	ISBN 1-882256-36-0
DODGE TRUCKS 1948-1960	ISBN 1-882256-37-9
LOGGING TRUCKS 1915-1970	ISBN 1-882256-59-X
*MACK® MODEL AB**	ISBN 1-882256-18-2
*MACK AP SUPER-DUTY TRUCKS 1926-1938**	ISBN 1-882256-54-9
*MACK MODEL B 1953-1966 VOLUME 1**	ISBN 1-882256-19-0
*MACK MODEL B 1953-1966 VOLUME 2**	ISBN 1-882256-34-4
*MACK EB-EC-ED-EE-EF-EG-DE 1936-1951**	ISBN 1-882256-29-8
*MACK EH-EJ-EM-EQ-ER-ES 1936-1950**	ISBN 1-882256-39-5
*MACK FC-FCSW-NW 1936-1947**	ISBN 1-882256-28-X
*MACK FG-FH-FJ-FK-FN-FP-FT-FW 1937-1950**	ISBN 1-882256-35-2
*MACK LF-LH-LJ-LM-LT 1940-1956 **	ISBN 1-882256-38-7
*MACK MODEL B FIRE TRUCKS 1954-1966**	ISBN 1-882256-62-X
*MACK MODEL CF FIRE TRUCKS 1967-1981**	ISBN 1-882256-63-8
STUDEBAKER TRUCKS 1927-1940	ISBN 1-882256-40-9
STUDEBAKER TRUCKS 1941-1964	ISBN 1-882256-41-7

The Iconografix Photo Album Series includes:

CORVETTE PROTOTYPES & SHOW CARS	ISBN 1-882256-77-8
LOLA RACE CARS 1962-1990	ISBN 1-882256-73-5
McLAREN RACE CARS 1965-1996	ISBN 1-882256-74-3

The Iconografix Photo Gallery Series includes:

CATERPILLAR PHOTO GALLERY	ISBN 1-882256-70-0

All Iconografix books are available from direct mail specialty book dealers and bookstores worldwide, or can be ordered from the publisher. For book trade and distribution information or to add your name to our mailing list contact

Iconografix
PO Box 446
Hudson, Wisconsin, 54016

Telephone: (715) 381-9755
(800) 289-3504 (USA)
Fax: (715) 381-9756